THE BOOK FOR MATH EMPOWERMENT

RETHINKING THE SUBJECT OF MATHEMATICS

D0110737

THE BOOK FOR MATH EMPOWERMENT

RETHINKING THE SUBJECT OF MATHEMATICS

Sandra Manigault

GODOSAN PUBLICATIONS

Published by
Godosan Publications
P.O. Box 3267
Stafford, Virginia 22555

Library of Congress Catalog Card Number: 97-072334

ISBN 0-9658541-0-8

Cover art by Donald A. Manigault

Cover design by Janeen T. Jackson

Printed in the United States of America

10 9 8 7 6 5 4 3 2

*This book is dedicated to
my father, Silas L. Young, and to the
memory of my mother,
Jacqueline Young.*

ACKNOWLEDGEMENTS

I wish to thank my husband, Donald, for his loving support and technical assistance during all the months of writing, and for gracing the cover with his beautiful art; my son, Patrick, for his editorial assistance and timely encouragement; my daughter, Dawn, for teaching me everything I know about the MacIntosh computer, her editorial insights, and for being the perfect role model; Mama Manigault for always believing in me; Nancy (Ayanna) Wyatt for her patient editing throughout; Cecilia White for her refinement of the final draft; Don Goral for his assistance in creating the graphics; Janeen Jackson for a beautiful cover design; B.J. Plenty for her excellent photography; Dr. Richard Semmler for giving me a schedule that allowed me to write; the wonderful staff at Porter Library for their outstanding generosity and friendly assistance over many months; Julia Cameron for the inspiration I found in her fabulous book, <u>The Artist's Way</u>; and my many students in math 120 ----- you were my initial inspiration for this book.

Contents

****

****

THE BOOK FOR MATH EMPOWERMENT

RETHINKING THE SUBJECT OF MATHEMATICS

THE BOOK FOR MATH EMPOWERMENT
RETHINKING THE SUBJECT OF MATHEMATICS

Introduction

Have you ever felt uncomfortable in a math class, nagged by the belief that everyone but you understood what was taking place? Do you sense that you could have done more with your life professionally and financially had you a better working relationship with math? Did you grow up afraid of math or insecure about your abilities because of it? If you answered "yes" to any of these questions, you are not alone in your perceptions.   There are probably millions of people who are math inhibited, math scared, or math scarred. This book has been written for all of you: those with a history of negative math experiences who wish to   feel intellectually whole again; those who wish to experience math recovery and math empowerment; and those who wish to provide positive math experiences for their children or their students. The purpose of this book is to provide a pathway to healing, while at the same time to shed light upon how to learn a much misunderstood subject.

There is so much hype about the subject of mathematics in American culture, and part of the hype is that some people are not math gifted. This book is not for those who love the subject or excel at it. It is for everyone else. This book is intended to provide an anecdotal, motivational, and metaphysical approach to a subject much of the population has learned to dislike. Unusual topics will be discussed, such as the relationship between learning math and music, false beliefs and their negative impact, the metaphysical connection, the significance of the right teacher, reprogramming the subconscious to accept the subject through affirmation, and much more.

The affirmations in the opening chapter are

for you to use <u>on a daily basis</u>. There is nothing unrealistic about this approach, and I do encourage you to try it for the duration of any math course that you may be taking. What is most important is that you read the book. Swallow it whole and come back to chew upon parts with which you may disagree. I wish you joy as you look with another view upon this subject.

******

PART ONE

TOOLS FOR

PERSONAL
EMPOWERMENT

CHAPTER ONE

The Purpose of Affirmations

As we think, so we are. Negative thinking restricts and destroys productivity. That part of us which leans toward the sunlight of possibility is that part which learns. Learning occurs where hope exists and asserts itself.

If we exist in a realm of possibility, we allow ourselves the option of succeeding. Affirmations are a powerful form of "self talk." Affirmations, used over a significant period of time, change our beliefs, allowing us to function more effectively in many areas. Affirmations are empowering. They allow us to believe we are

5

wonderful beings, capable of learning and succeeding at what we attempt to do.

Affirmations help us to clear blockages from past turbulence. They proclaim a new threshold from which our dreams are born. Like shining a foggy mirror, affirmations help us to see ourselves and our potential more clearly. They allow us to firmly believe that we possess power, intelligence, skills, and resources. They help us believe that we can begin to do things that seemed impossible. Affirmations throw open the gate to our success.

The affirmations in chapter one are designed to address the question of "Can I do math?" and to be the foundational whip for the answer. They are to be a buffer from past mislearning and the salve for healing emotional scar tissue. People who dislike math have deep seated reasons for their aversions. The affirmations forge new bridges back to the subject.

People may not always admit the truth of their feelings, not even to themselves. However, the subconscious cannot be fooled. Deeply embedded fears block learning, and deeply entrenched negative beliefs obstruct progress.

To assess oneself fairly and gently is important. This is, in part, what the affirmations attempt to do: to assess correctly both oneself and the subject of mathematics.

There are a lot of false beliefs about math. We address one of these in the affirmation: "Long does not mean the same as difficult." Recovering from our false beliefs is one of the goals of chapter one. As we begin to think about what is true and what is not, we are in more of a position to attempt strategies others have found to be successful. Believing those strategies have a chance to work for us will give us the courage to try them. We might even become successful for the first time.

******

AFFIRMATIONS FOR LEARNING MATHEMATICS

1. Everything I need to know comes to me when I need it.

2. Each day math becomes easier to understand and more satisfying to me.

3. The classroom environment is comfortable and safe. I am free to learn in any way that I wish.

4. There is no ridicule or fear. I am at ease and happy in this environment.

5. My memory is good, and I retain a perfect understanding of all the concepts.

6. My logic mind perfectly absorbs all definitions, formulas, rules, and theorems.

7. My creative-visual mind grasps and remembers all concepts and diagrams.

8. My brain is capable of integrating and applying all that I see and hear.

9. I am fully capable of achieving any grade that I wish and am willing to work for. I have the capacity for success.

SOME TRUTHS
ABOUT MATH AND ME

(Reprogramming the subconscious to do math with less
resistance)

1. I am meant to succeed in mathematics.

2. I am meant to do math easily.

3. Math is easy at this level and I can do it well.

4. "Long" does not mean the same as difficult.

5. I can do anything I put to my mind.

6. Practice and persistence are the keys to
    success for me.

7. Algebra is fun and algebra is easy.

8. My mind is systematic and analytical.

9. Formulas are easy to memorize.

10. I do not choose to procrastinate.

11. The learning of mathematics enhances my
    other creative efforts.

12. It is easy to work successfully in small
    blocks of time.

13. I can depend on myself to get things right.

14. I am creative, resourceful, and intelligent.

MAKING MATH EASIER---- TRAITS AND BEHAVIORS THAT WORK

TRAITS

*a positive attitude   open-mindedness   consistency
patience   persistence   ability to stay focused
desire   precision   commitment to preparation
willingness to learn concentration   the setting of goals
adherence to schedules   belief in one's own intelligence
commitment to homework   recognizing the need to
improve   self-confidence   optimism*

BEHAVIORS

*paying attention to detail   asking questions   reviewing
materials   studying regularly   seeing math as
more than numbers   listening   avoiding self-deceptions
reading for comprehension   limiting distractions
not skipping steps   alternating approaches
taking study breaks   looking for practical applications
using positive self-reinforcement   devoting time to the task
mastering the basics   visualizing the steps
setting priorities   the discipline to open the book*

Math Affirmation Series
Copyright 1995   *Sandra Manigault*

NEW PARADIGMS

*IN CONDITIONING MYSELF TO FORGE A NEW
PATH, I CHOOSE NEW POSTULATES BY WHICH
TO DIRECT MY LIFE. I CHOOSE TO BE*

PATIENT
> Patience is a form of faith that allows me
> to believe in what is possible.

PRACTICAL
> I recognize that work will be required if I
> am to make significant progress.

PERSISTENT
> If I quit before I succeed, then my
> success cannot be born. I cannot abandon
> something merely because getting it is
> difficult.

PRECISE
> In this era there is no room for shoddiness
> and carelessness. I will commit to doing
> things well.

POSITIVE
>   My attitude governs how I will be on a given day. To succeed I need to believe both in myself and in what I am doing.

PRACTICE-ORIENTED
>   I cannot master what I do not do.

PEACEFUL
>   In serenity I will connect with my Higher Self and tap the potential that lies within me.

****

AFFIRMATIONS OF DENIAL

1. I do not allow myself to be intimidated by the negative energies of others.

2. There is no need to feel threatened. I am as smart and wonderful as everyone else.

3. I am not calculator dependent. I compute quickly on paper or in my head, and I am correct.

4. I can learn regardless of who is teaching.

5. I do not hang onto painful experiences of the past or present. As I release these, I am free to learn and grow.

6. There is no need to feel competitive. Each person wants exactly what I do --- health, love, and happiness.

7. I do not fear mathematics. It is a part of life.

AFFIRMATIONS OF EMPOWERMENT

I am not a victim of the past or present.

I am not hurt by the actions or choices of others.

I choose how to spend my time and energy.

I choose to be healthy, wise, creative, and free.

I am psychologically uninjured.

I choose the circumstances of my life.

I choose the life I want to live.

I choose to do what is good for me.

I am not governed by guilt.

I choose to have peace of mind.

I choose to be whole, happy, and free.

AFFIRMATION FOR TOTAL HEALING

I AM A CHILD OF GOD.

I AM FILLED TO OVERFLOWING WITH GOD'S HEALING POWER.

I AM COMPLETELY HEALED IN MIND, BODY, SOUL, SPIRIT, THOUGHTS, ATTITUDES, AND EMOTIONS.

I AM COMPLETELY HEALED IN THE SOLAR PLEXUS CENTER.

I AM COMPLETELY HEALED IN THE HEART CENTER.

I AM WHOLE. I AM HAPPY. I AM FREE.

THANK GOD. THANK GOD. THANK GOD.

AMEN.

CHAPTER TWO

*Study Skills for Mathematics....Do's and Don'ts
From Real Situations*

In the case studies that follow, we are going to expose some errors in thinking and behavior that have caused students to fail. While the intention is not to belittle anyone's personal situation, it is important that we learn to identify attitudes that make it impossible to succeed. In my years as a community college teacher, I have witnessed all modes of conduct, ranging from immature adult behavior to the assuming of more responsibility than is manageable. While many of our students did not

18

take full advantage of their educational opportunities in high school, a greater number seem to lack coping skills. They appear to be overwhelmed with personal problems that make getting an education very difficult.

Anecdote 1

Bob was enrolled in developmental math, an individualized course designed to allow him to begin wherever he was, which for Bob meant with the basics. Even with the assistance of a calculator, Bob had trouble. He lacked patience. He also lacked an appreciation for exactness. However, in the context of a professional environment, he could understand the need for perfection where the exchange of money for personal services were involved. When the idea of mathematics was couched in this context, he could better relate to how his own attitudes were in his way. (In this era there is no room for shoddiness and carelessness).

****

Anecdote 2

Jennifer's behavior was gregarious and bubbly in the math lab. When we discussed her first test, on which she received 20%, it was as if she never stopped socializing long enough to learn any mathematics. "But, I go blank whenever I take a test," was her justification. Hidden in that comment were the words "I can't do anything about it, so how are you going to compensate me?" The ball went back. "How then are you planning to get through college?" In three minutes it was apparent that she was not only ignoring her text, but was relying on what she remembered from high school. After an explanation of how to do her test correctly, she was gently reminded that studying and rule learning would be required to fix the "blanking out." She was also expected to stop socializing and do her work. (Work is required if one is to make significant progress).

Anecdote 3

Yasmin was a bright student enrolled in pre-calculus. She could answer nearly every question in class but failed to get more than a "C" on tests. The discrepancy was startling. Closer examination indicated that she was not studying for tests sufficiently. Looking over notes the night before formed the extent of her preparation. There had been no rereading of the text, no review of homework problems, no sample problem solving. A more in-depth approach to studying improved her grades. (One cannot master what one does not do).

Anecdote 4

Doris was enrolled in algebra. Although in attendance each day, she was not passing. She sat quietly taking her notes and failing her tests. After observing this behavior for three weeks, we talked. She was a young, single mother of a rambunctious three year old. She had excellent child care while she attended college, but failed to avail herself of this service while at home studying. Her child was getting the best part of her time and energy. I explained the need to work in the library before class or after class, extending her school day. In this way she could give her undivided attention to herself and her child, without conflict.

Anecdote 5

Douglas was a serious young man of 20, enrolled in pre-calculus with me for the second time. He failed still. Recognizing the pattern, I asked if we could talk. He was working 60-80 hours a week. I lost my discretionary space, and the questions flew. Why? Why was he overworking? Did he have a wife and three children to support? Was he living beyond his means? To work such long hours while still living at home with his parents was ridiculous. Although he seemed to understand that getting a college education was full time work, he did little to alter his schedule. He withdrew from class a second time.

Anecdote 6

Steve was taking a night class after work and was failing miserably. On probing to learn when and where he managed to do his studying, I learned that after work he would go home to run errands for his wife before coming to class. In class four nights each week, he had very little time to study. I pointed out the folly of this arrangement and suggested he partake of fast food after work and head to the library before class. Although he acknowledged the logic in having his wife run her own errands during the day, nothing changed, and he failed.

Anecdote 7

Pamela was one of my most conscientious algebra students. Married, in her 30's, she worked hard and did marvelously. It was shocking to find her at my desk, in tears, toward the end of the semester. Her husband was opposed to her continuing in her nursing program. She kept his house, cooked his meals, but he resented the time she spent studying in the evenings. Between tears, she asked me what she should do. I advised her to stay in school. If she felt a calling to save lives as a nurse, I felt that was what she ought to do. I went on to say that one of those lives might even be her husband's.

(In today's tight job market, few careers offer as many options as nursing. Furthermore, many people marry before completing their education. Spouses need to respect this need to go back to school when it occurs.)

****

PRIORITIES --- A REALITY CHECK

EVERYTHING IS NOT EQUALLY IMPORTANT.
Succeeding, in part, is an issue of intelligent
prioritizing: putting long range goals first.
Succeeding compels students to differentiate
between what is important and what is urgent, to
examine how time is spent, and to determine
how choices are made.   Students must ask
whether ample amounts of time are being spent
addressing what is really important (ie., getting
a degree), or are disproportionate amounts of
time being wasted? Do we allow feelings of guilt
and obligation to compel us to put the agendas of
others before our own?

Some of us are propelled by our own
insecurities, and like Jennifer, who over
socialized, or Douglas, who overworked, we
become totally unrealistic. Lacking vision and
commitment, we have difficulty saying "NO" to
all the distractions that jump in our path.

In recent classes, I have asked my students
to tell me, in writing, what is on their minds.
These brief "emptyings" reveal where their
energy is being directed and the nature of their

priorities. I remember the paper of a young man, torn between being a good son to his mother (in her own crisis), a good friend, a reliable employee, and a good student. He never made it to his last goal. Until we learn to keep our agendas and set our problems aside, we will not reach our long range vision.

PROBLEMS --- A SPIRITUAL VIEW

From another perspective, we might ask from where do all of life's crises come? Are they here to torment and distract us, or do they serve some higher purpose? To what extent is a pattern of helplessness on our part self-imposed? Is resolving a problem a matter of focusing our attention elsewhere, rising above our emotions, or just letting go? How do we quiet nagging worries long enough to accomplish something of value? Should we schedule in daily time for self, a time in which the priority is to listen to ourselves, to nurture our own spirits? Should we cap the quantity of time we allow ourselves to "give away" to others, and where in the scheme of time do we draw the proverbial line?

We need time for ourselves. This recognition is spiritual in that it acknowledges our need to reconstitute ourselves after the living we do has depleted us of that which makes us "go." This time for renewal, whether it occurs in meditation, quiet walks, or in some other form, allows us to put issues in perspective and get our lives in step with our Higher Self or Creator. When we start to come "unglued," it is because we have denied ourselves this basic necessity.

EXERCISE

1. Go back to the paragraph of questions and write out some answers.

2. Identify 5 ways in which you "reconstitute yourself."

****

AN AGREEMENT WITH MYSELF

To become successful I must choose to face issues I have avoided. I must choose to let go those things which have held me hostage. To grow, to expand, I must release those beliefs that have corrupted my consciousness. I must relinquish negative habits of overstimulation, overfeeding, numbing myself (with deafening music or mindless television), and other habits that destroy my sensitivity. I must reroute energies and reprogram thinking. I must stop letting others dictate how I should live and whom I should be. I must free myself to be myself. I must get out of my own way. And I must get out of the way of others. I must let go of fruitless and unsatisfying relationships. I must learn to accept what I cannot change. I must stop camouflaging what I really want with empty activities. I must take responsibility for my life. I must choose to grow up. To be a success, I must adopt those lifestyle rituals that make others successful--- being decisive, focused, organized, realistic, determined, hard-working, patient, and serene.

****

AN EXERCISE --- MAKING CHOICES

1. Susanne takes two community college classes in the morning, works four hours in the afternoon, picks up her son from day care after work, and spends her evenings with her family while studying. Today she gets a call at work. Her son's temperature is 102 degrees. A dinner guest is expected that evening, and a test, for which she is unprepared, is scheduled for tomorrow. What are the priorities here? How do YOU rank them in terms of urgency and importance, and how would you cope in her situation?

2. Joy needs to buy a wedding gift for a good friend. After 30 minutes of shopping nothing within her budget emerges. Should she continue to shop or find another remedy?

3. Joseph gets a distressing call from his best friend who has just broken up with his girlfriend and needs to talk. It is 8 PM, and Joseph has a final exam tomorrow. What should he do?

4. Cecilia has a 3.92 GPA and is struggling not to loose it to a "C" in physics. Her physics

professor has assigned two more labs and a quiz, all due within three days. She also has a research paper to complete for English. What should she do? Is the best remedy one of "letting go" at this point? In her case, would letting go be the same as giving up?

5. Think of a recent crisis you have experienced. How would you handle it today?

STUDYING

When burdened with some extraneous issue, studying anything is difficult. While controlling outcomes is not always possible, relinquishing *the need for control* and for closure are. Attempting to restrain the need to worry becomes a critical factor in governing the quality of our lives. Students who succeed are focused on what they are studying and have managed to achieve a "rise above it" attitude. They either minimize their stresses or ignore them, in that they do not allow stresses to rule their emotions. They learn how to function in a worry-free state, recognizing that *emotional control is key to mental control* --- the foundation of academic achievement. *The mind can only do what we allow it.* Scattered thinking and displaced emotions bar the door to control and, thus success. We must learn how to maintain focus.

Think of the mind as a vast array of light, like a burst of sunlight. It is very powerful, but diffused. To learn with optimum speed and accuracy, the mind must be focused like a single laser beam of energy. All extraneous factors must be pushed aside.

Besides working in a worry-free state, studying in small blocks of time works well and minimizes stress. Making lists of what needs to be done, alternating between silent and aloud reading, reading with pencil and paper, reading with a yellow marker, alternating reading and analytical subjects---all these variations keep the mind alert and cooperative. Learn to trust your instincts. Doing what feels right at the moment gives you both freedom and flexibility. While it is necessary to follow some kind of schedule, avoid making a tyranny of the "shoulds."

Learn to keep bothersome people at a

distance to protect your space and moods. Don't let others waste your time and cause you to loose your focus. Your study time must be protected and utilized if you are to succeed. Study when and where you are unlikely to be interrupted.

Unlike most subjects, mathematics needs to be read repeatedly. The failure to do this is why so many students are always saying they don't understand. *Mathematics is concept-laden, not fact-driven,* which is why a single reading is so insufficient. Classroom note-taking becomes a futile effort unless notes are used to summarize and reconstruct the lesson. Classroom notes should be swallowed whole once weekly, and reviewed as necessary during the week, to facilitate doing the homework. Your notebook traces the priorities of your teacher, whereas the text follows a more generalized curriculum. Reading both is important.

Test preparation occurs best over a three day period. This maximizes your opportunity to read the text, review your in-class notes, and practice sample test questions found in your assignments and textbook chapter reviews. While all of this sounds like too much work, think of it as a redistribution of your time ---

time you previously misspent. The best way I know to get an "A" is to study the material until it seems trivial. But, lacking this luxury, concepts and formulas need to be memorized, and mechanical operations practiced until they become automatic. This allows you to focus on the conceptual aspects of the test while avoiding most careless errors. Finally, when taking a test, take the time to proof-read every solution since the majority of test errors result from carelessness. Allow patience and precision to take charge and work magic.

I'M IN THE MOOD FOR ---

While few people would finish this sentence with "math," let's move beyond the limiting belief that "I can't study math until I'm in the right mood." Let's recognize that mood improves with invigorating exercise, diversified activity, or stimulating music. A longed-for silence or a roaring fireplace adds much needed comfort to the scenario. Improve your working environment by identifying distracting factors and removing them wherever possible. How much easier the "mood" is to achieve once you have made your education the priority.

You will find it easier to work on mathematics in intervals of up to one hour, restricting boredom and tension. It is best to follow this math hour with a radically different subject like French or writing. This will rest the math mind and use a different cognitive process. Rotate subjects, take reasonable breaks, and study early. Bored, tired minds don't work well.

The following is a list I give each class with its syllabus. Please use it. It will help you manifest your goals.

SOME HELPFUL SUGGESTIONS

1. Be sure to come to <u>all</u> class meetings. Problems occur when continuity is disrupted due to absenteeism.

2. In addition to taking notes in class, it is a good idea to read those notes before doing homework assignments.

3. It is important to read your textbooks several times each week. This strongly reinforces whatever is covered in class.

4. Test preparation is critical. Preparation should begin about three days before each test, or occur in three different preparation periods. Preparation for the final exam should be much more extensive than that which is required for a chapter test.

5. Please make use of my office hours. If you need help or wish to see me and these hours are not convenient for you, please see me during class, and I will make other arrangements to work with you.

6. If it is necessary for you to be absent, please be sure to call me or a fellow student to find out what was covered and to get the homework assignment.

7. Ask questions when you have them, but please be considerate of your classmates. Make a point to get to class on time. A great deal of important material is missed by habitual latecomers.

****

MYTH VS. MASTERY

Three weeks into a new semester, I inquired of my introductory math class, "How many of you have learned the rules for fractions and integers?" One lonely hand rose. I could not believe it, as I had delivered my best speeches only days before. Consequently, I permitted myself to be sarcastic: "What are the rest of you waiting for--- show time?"

Sometimes muddled scheduling, sometimes a lack of realism creates unnecessary pockets of failure in our experience. Unwisely, students believe "success happens." It does not. It is deliberate: conceived, visualized, planned, executed, and reworked. In math, seeing is not doing; hearing is not learning. We do not enter a new situation with all of the answers. We should expect to expose and heal our flaws as we learn. *If we hear good advice, but fail to understand the consequences of what is being said, we may forfeit our ability to act intelligently.* We will be like the actor who has memorized his lines the night before. We will flounder and act foolishly.

If we try to cram a math chapter as if it were history, we are failing to distinguish the learning of concepts from the learning of facts. *Concepts don't cram.* Concepts must be understood before they can be memorized, practiced, mastered, applied. The faked test will eventually yield a "D" and we may falsely conclude that the process is not working. We will have failed to get the point, which is that *math courses won't cram.* They are concept driven, not fact based. At multiple levels, these concepts are presented visually, verbally, and symbolically. There must be the time to correlate and synthesize. We must read and ask ourselves at what level do we <u>not</u> understand? Visual? (What does this diagram mean?) Verbal? (What does this explanation say?) Symbolic? (How do I apply these formulas?) What connection has not been made and where can I get clarification? We must put time into mathematics, time that we have set aside for ourselves.

There are subtle nuances in how we read math, write math, and reason in math. Understanding the subject requires a slowing down of the learning process for careful attention to detail. Success in math requires

that we do it the right way---completely.
Homework cannot be short cut. Processes cannot
be glossed over. Details cannot be ignored.
Admonitions to study cannot be trivialized. We
have to put our act together and commit to
mastery. Then we can do it. Then we will
succeed.

STUDY SKILLS SUMMARY

DON'TS

1. AVOID letting others consume time you
   have allocated for studying.

2. AVOID studying after 9 PM.

3. NEVER take a test tired or without
   studying for it.

4. DO NOT study collaboratively if you have
   not reviewed on your own.

5. AVOID getting behind. Catching up in math
   is difficult.

6. AVOID studying in long blocks of more than

1 1/2 hours. Change subjects or take a break.

7. AVOID self-deceptions. Acknowledge what you do not know.

DO'S

1. BECOME your first priority.

2. STUDY during prime time: 8 AM to 8 PM.

3. READ your affirmations and class notes before doing homework.

4. ALLOCATE more time for test prep than you think is necessary.

5. ASK questions when you have them.

6. FORM study groups whenever possible and expedient.

7. MEMORIZE formulas, definitions, and processes when they are initially presented.

EXERCISES

Think about and respond to the following.

1. We are all infantile in our emotions in those areas of our lives in which we fail. The question is --- how do we grow up?

2. Each of us is responsible for our own successes, both academic and personal.

3. A commitment to practice is what makes us good at something.

4. We should not confuse being smart with being committed. *We are already smart.* Successful people are committed.

5. Loving ourselves means to stop looking for something to criticize.

6. I do not have to impress others. I have only to know my own power.

****

CHAPTER THREE

"REAL LIFE" APPLICATIONS FROM A METAPHYSICAL STANDPOINT

All of life is a risk, whether to go white-water rafting or take algebra after 20 years. What does the older student risk? There is ridicule: confirmation of one's greatest fear, that one is indeed stupid. There is the risk of not keeping up with "kids" out of high school, or the risk of non-acceptance by a fast talking teacher. There is the risk of shame ---"I can't do it."

But the above are outcries of the ego, which reflect false assumptions. The real you, the *essence* of you, wants to do the math because it knows YOU CAN DO IT. The real you has no fear because it already knows everything. By definition, education is a "drawing out of." All knowledge must be brought to a conscious awareness level if you are to make use of it. It is that effort which takes the time, requiring the "hard work."

Preparing to succeed can start as a simple inner exercise in visualization. For fifteen minutes sit quietly, breathe deeply, and quietly imagine a perfect class in which you arrive early and seat yourself comfortably. Imagine the instructor entering with a warm smile and teaching an invigorating lesson in which you understand everything. See yourself interacting positively with the teacher and the students. See yourself happy in that environment. Doing this regularly will help you allay your fears and become more comfortable.

At home, create a mood for *math*. Play some soothing, wordless music (ocean waves, pan flute, etc.), or some baroque music (Bach or Mozart). Find a quiet, attractive work station, and work uninterrupted for one-hour intervals, stopping only if anxious or tired. Before beginning, imagine yourself understanding the text and remembering the day's lesson.

A third and powerful approach is to engage in positive talk with yourself, using your chapter one affirmations. These will push you past any unpleasant blocks and open channels for easier absorption of the material.

MISTAKES IN THINKING

STUDENTS HARM THEMSELVES BY COMPARING THEMSELVES WITH EACH OTHER. We are each unique entities with a different array of potentialities and purposes, different in energies and directions. Each of us is like a unique version of the linear equation $y = mx$.

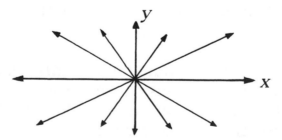

As each equation has a different slope (slant), each life has a different focus or direction. Nothing we accomplish will duplicate another's efforts. Comparing our logical-mathematical abilities is as foolish as comparing our voices. Students need to relax and let go of the need to compete, recognizing the number at the top of a test paper does not evaluate total ability. We need to also let go of the false belief that she who finishes first wins (a belief left over from elementary school conditioning). Isn't it strange, in terms of life span, we come to believe just the opposite: that the winner is she who finishes last.

HOW DO I STUDY IN THE MIDST OF PAIN?

Although all of us are at this juncture at some time, emotional pain can be overwhelming. But life ignores us in her unending ride through time, as we wake up from our malaise weeks behind schedule and disoriented. Most of us are not aware that we *choose* to be disengaged from life in our wallowing, that we choose the emotions we feel in much the same way we choose what shoes to wear; and, after we put on debilitating emotions and attitudes, we refuse to shed them for they serve a purpose. They feed our fear, our stubbornness, our self-pity. In our tenacity, we either refuse to stop dallying with them, or we refuse to get the spiritual help we need to overcome our problems.

Most believe in a Higher Power. But, how many of us ask to be "released from" that which is tormenting us? What is it we are trying to avoid---the work or the pain? Is our negative preoccupation just another form of procrastination, or are we indeed helpless? Studying is a way to refocus our attention, and we should agree that attending class could be a blessing. So many students have withdrawn from classes or failed because of problems they could not generate the strength to overcome. In dealing with problems and personal pain, it is a good idea to meditate, do affirmations, make lists, follow schedules, eat well, exercise, and look good.

The following is a powerful affirmation (repeated from chapter one) designed to heal the physical, mental, or emotional body. You may wish to commit it to memory.

I AM A CHILD OF GOD
I AM FILLED TO OVERFLOWING WITH GOD'S
    HEALING POWER.
I AM COMPLETELY HEALED IN MIND, BODY,
    SOUL, SPIRIT, THOUGHTS, ATTITUDES,
    AND EMOTIONS.
I AM COMPLETELY HEALED IN THE SOLAR
    PLEXUS CENTER.
I AM COMPLETELY HEALED IN THE HEART
    CENTER.
I AM WHOLE. I AM HEALED. I AM FREE.
THANK GOD. THANK GOD. THANK GOD.

When all of the above has been accomplished, do something pleasant. Visit a park, take a long walk, or go to the movies. Escape for a day. Then return to your life revitalized and ready to get on with it.

METAPHYSICAL MATHEMATICS

Every contemporary math text contains real life applications of relevant math concepts. This idea is commonplace. What I have attempted to do in the pages that follow is somewhat different.

I have used graphs to illustrate some relational aspects of our experience and looked for new illustrations for old concepts. The connection is "metaphysical," not physical in the sense we see it. I hope you will see the connection between the math concept and the spiritual tenet associated with it. For some of you this will be a reach. Nonetheless, it will give you a new context for reexamining some familiar concepts.

RELATIONSHIPS

Every relationship has its own special purpose and pattern. Another way to appreciate mathematics more is to use graphs to illustrate some human connections.

******

THE HARMONIOUS PARENT-CHILD RELATIONSHIP
(Working together to achieve common goals)

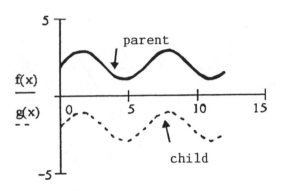

THE BRIEF AFFAIR
(One connection)

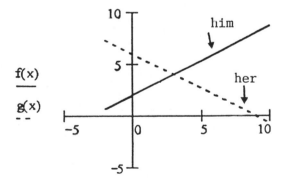

****

THE MENTOR-STUDENT RELATIONSHIP
(The student may flounder, but the mentor's support is constant)

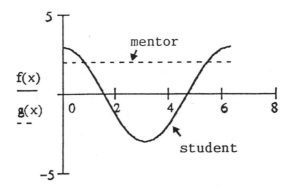

THE SUCCESSFULLY RECOVERING DRUG ADDICT
(A step function even in mathematics)

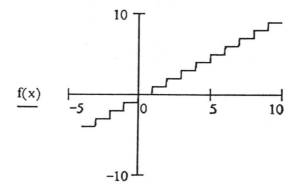

****

SOUL MATES

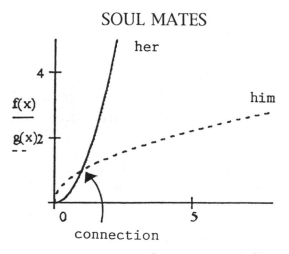

In this case, $f(x) = x^2$, $g(x) = x^{1/2}$ and $f(x)*g(x) = x$. Their composition yields the identity function.

A METAPHYSICAL LOOK AT SOME MATHEMATICAL BASICS

On repeating decimals---some people are just like these decimals: they repeat patterns endlessly because they won't learn how to get it right the first time.

$$1/3 = 0.333333... \qquad 23/99 = 0.232323...$$
$$1/9 = 0.111111... \qquad 123/999 = 0.123123...$$

On the number zero--- let us consider zero as the beginning of life, the integers before zero representing "pre-birth," and those after zero, life in years, as we know it.

$$\frac{f(x)}{\quad} \quad \overset{\longleftarrow\!\!\!|\!+\!|\!+\!|\!+\!|\!+\!|\!+\!|\!+\!|\!+\!|\!+\!|\!\longrightarrow}{\underset{-5\,-4\,-3\,-2\,-1\;\;0\;\;1\;\;2\;\;3\;\;4\;\;5}{}}$$

The infinity symbols do not represent real (decimal) numbers. They represent unboundedness and limitlessness. Whereas the interval (- ∞ , + ∞) represents the entire real number line, it may also symbolize the unboundedness of life, the path of the soul, without beginning, without end.

The ABSOLUTE VALUE of a real number is its distance from the number zero on the number line. It is a non-negative quantity, that is, the absolute value of a number cannot be negative.

$$|x| \geq 0$$

****

Metaphysically speaking, perhaps the absolute value of a life is the amount of soul growth that has taken place therein. As absolute value cannot be negative, every life serves some purpose in terms of soul growth. A life that *appears* to be negative or valueless in reality may accomplish just as much in terms of soul growth as one that appears to us as positive. If we look again at the geometric definition of absolute value, the distance of a number from zero on the number line, we are seeing how much has been accomplished in the soul of a life from birth.

ALGEBRA

It is characteristic of algebra to use letters we call variables to represent real numbers, that is, all decimal numbers. This symbolic depiction allows us to talk in generalities and draw broad conclusions instead of citing specific example after example to prove a point. The inclusion of many definitions and refinements tends to overwhelm a lot of students, inducing them to believe all of this constitutes difficulty. However, this is not true. The definitions are there to help us remember what we are doing by supporting the explanations. On the real numbers, we have four basic operations---addition, subtraction, multiplication, and division. Although addition and multiplication may be construed as the primary operations, in fact, all emanates from addition alone. We are often shown multiplication as "accelerated" addition.

For instance, $2+2+2+2+2 = 5(2)$, so defines the multiplication of 5 by 2.
$2 - 5 = 2 + (-5)$, so defines subtraction as the inverse of addition.
$2/5 = 2(1/5)$, so defines division of 2 by 5.

At a higher level we see in $2(5^2)$ the exponent 2 indicating accelerated multiplication, that is, $2(5)(5)$. A closer look reveals that $2(5^2) = 2(5)(5)$, or $5(5) + 5(5)$, or $(5+5+5+5+5) + (5+5+5+5+5)$. ULTIMATELY, all of this becomes $1+1+1+1+...+1$ for 50 ones, which leads us to one final insight. All numbers are generated from a single element---the number one. All of life, likewise, emanates from a single source, God, and is translated into an infinite variety of forms and vehicles.

BEFRIENDING THE RULES

All of those rectangular bordered passages, color-coded messages, italicized symbolic phrases that grab our attention in textbooks need to be identified as "friends." These formulas and definitions are guides to help us sort our way through unfamiliar terrain, to bring structure to an otherwise indecipherable medley of number-symbols. They tell us what to do and how to keep on doing it. In addition to befriending these rules, we must pay attention to caveats of a different nature, the "restrictions," such as "division by zero is undefined." Mathematically speaking, if a and b are real numbers, a/b exists provided b does not equal zero. From a METAPHYSICAL standpoint this restriction serves a major purpose: it tells us ALL MUST BE SUPPORTED BY THE LIFEFORCE. Without the support of the lifeforce (God), a "thing" is dead, hence, does not exist.

Algebra simulates life in other ways. There exist multiple ways to solve problems. It is a lack of creative thinking or a lack of patience that forces us to attempt to settle all issues using a singular strategy. As we read collections of rules and properties in the beginning of a chapter, we are being readied for what is possible. In mathematics, as in life, there is the element of surprise. The rules prepare us for all contingencies, as well as lend structure to chaos.

With so many rules structuring the implementation of mathematics, why are so many students confused about what to do, and why are so many making careless errors? To stay "above the law" one must follow the rules to the "letter of the law." Mathematics can be understood at the conceptual level, like driving by a forest woodland at 35 miles per hour, or at the specific-detail level, like taking a solo hike through it. How much more one can see and appreciate on foot; how much richer the experience and how much better it is remembered. No one recalls much detail on a 35 mile per hour drive through.

TRUSTING THE UNIVERSE

Have you ever been "closed out" of a class, routed to another, and found the new teacher "better" than the one you would have had? Has a traffic jam ever protected you from an unfortunate accident? Were you denied admission to a particular college only to later meet your spouse at your "second choice?" All of these "coincidences," are they not divine guidance at work in your life, like some unseen hand working out your life's details better than you could have?

We need to arrive in our thinking at that juncture in which we put our faith in God, trust the direction in which we are being "led," and accept that things which are going wrong are often the outworking of life's circumstances in a better way. This brand of intuitive awareness will prevent us from becoming frustrated and angry when life takes an unexpected detour. In our pursuit of success, we should seek out serenity as that spiritual state in which we allow a divine force to chart a better direction for us. As we meditate on the "perfect outworking" for the choices we must make, we come to understand that being told NO points us to a different option, an unseen blessing. We begin to view a job layoff as the ultimate opportunity to return to college, or the broken relationship as the freedom to meet a better partner. As our perspectives change, we learn to see life's problems as life's opportunities.

EXERCISES

1. Unexpected, happy "coincidences" are referred to as synchronistic events. Can you identify five examples of synchronicity in your life in the past two years?

2. As you read passages in your math text, what unusual insights about life are you experiencing?

****

PART TWO

IDEAS FOR

GREATER AWARENESS

CHAPTER FOUR

FALSE PARADIGMS, ILLUSIONS, AND OTHER HANG UPS

FALSE PARADIGM I
You can't learn with a bad teacher

This would be a great excuse to fail if it were true for other subjects. *"I don't like the teacher"* frequently justifies a failed math course, but it should not. What exactly do these five words say to us? Do they say that the teacher----

a) assumes we already know the material
b) is unclear in giving explanations
c) talks too fast
d) is unprepared for class
e) talks down to us
f) is intimidating
g) doesn't want to answer our questions
h) all of the above?

If the teacher profiled above were instructing history or French one could get a "B" in spite of this exhaustive list of shortcomings. Overcoming such an insensitive, pompous instructor is what is meant by the affirmation "I am not intimidated by the negative energies of others." (See Affirmations of Denial in chapter 1.) One must toughen the outer core and use teachers like these to build resilience.

FALSE PARADIGM II
Only very bright people major in math

Some very bright people do major in math. Other very bright people major in English, art, dance, psychology, business, music, or anything else they choose. Some people selected math by default. That is, they felt inadequate to specialize in subjects requiring voluminous reading or writing, or they could not succeed in chemistry or engineering. In fact, at least one person I know extremely well felt it was less burdensome to read twenty pages of math as opposed to struggling through two hundred pages of research. (What you don't know!)

FALSE PARADIGM III
Algebra is an abstract subject and some people will never be able to do it.

Years ago, a middle school counselor tried to convince me that our daughter, Dawn, was not "ready" for algebra, because it was a very abstract course. Although I felt her agenda had a hidden component, it also fascinated me that people who don't teach math are quick to predict who cannot do it. People untrained to teach algebra are unqualified to judge its level of abstraction for someone else. Basic algebra is not an abstract course, and its mastery merely requires some specific strategies. (Dawn, incidentally, went on to get A's in math and an exceptionally high GPA in high school.)

Assuming the textbook writers do not obscure it, basic algebra is an easy to understand, logically formatted course. The presence of variables (letters) does not constitute abstraction. It is a course I frequently refer to as cookbook mathematics (ie., just follow the rules). Basically, it is a different application of addition, subtraction, multiplication, and division in each chapter.

Most students have difficulty in algebra only if their teachers: a) do not talk enough, b) do not write enough, and/or c) fail to repeat what was unclear from the day before. Necessary to assimilate this course are attention to detail, practice, patience, and precision. What algebra requires is that one grow up to professionalism. Neatness counts. Organization counts. The ability to follow sequential directions counts. It is not a matter of being smart. It is a matter of being disciplined. Assuming the latter, algebra is fun, and algebra is easy.

FALSE PARADIGM IV
Standardized tests are a fair measure of potential.

Standardized tests are very narrow instruments. They measure a very limited type of left-brain dominance. Often they are so awkwardly worded that whole meanings of questions are obscured. Hyped to be reliable, they are just the opposite and serve to stratify the population along socioeconomic and racial lines. "Normed" on upper middle class students, standardized tests are designed to work for specific segments of society. Students who test "average" on these biased instruments frequently move on to make significant academic strides beyond their "predicted" performance both in college and in graduate school.

What most people do not know about these instruments is that they are designed to very rigid specifications. Dozens of hours are spent designing wrong answer choices to simulate and capitalize upon all the intelligent mistakes good students are likely to make. They do not resemble textbook questions, or follow standard curricula. Higher cognitive levels of learning are being tested: the ability to draw inferences, to make applications and evaluations based upon what one has read; to differentiate between essential, extraneous, and insufficient information. What is being tested are much higher thinking skills than are generally taught in *basic* college preparatory courses. Math questions on standardized tests may be totally unlike textbook questions students see in geometry, algebra I and II. The SAT, in particular, includes truly abstract questions, many multiple step questions, and questions that are to some extent irrelevant to a student's ability to do statistics, or calculus, which are the courses most college bound students ultimately take. While not suggesting that parents rage at school boards to "upgrade the curricula" to better prepare students for the SAT, I would suggest that a few aggressive parents study the SAT and question its relevance to what their children are going to do with their lives.

ADULT LEARNERS IN MATHEMATICS

Returning to school after many years is a mixed joy. There is the exhilaration of "finally getting to do what I've always wanted," contradicted by the anxiety of being out of touch, archaic. Fear intrudes and nags us into believing we must work excessively to catch up with kids who never ceased to learn. We construct quite a frenzy! What we need is a preliminary course in remaining calm concurrent with another in defining realism.

What older learners do not know is that most of them are going to outperform their younger counterparts. They have more drive and more focus.

Nonetheless, it is amusing to see adult learners return to college after 20 years of academic vacation, heaving under the compulsion to get all A's. Obsessing over grades, they bring to college the stresses they left at work. Once required to please parents with their grades, they now endeavor to impress their own children. Ashamed not to do as well in algebra as a daughter, not to write compositions as well as a son, Mom competes with them. This is a sad commentary that tells us we do not value what is significant--- the opportunity to learn.

Let's create a scenario. Would you handle it this way?

Son: "Hey Mom, how's the algebra going?"
Mom: "It's fine. Good teacher, good book, nice people."
Son: "So what kinds of grades are you getting?" (Role reversal).
Mom: "B's and some C's." (Mom is now playing the child).

Son: "You're getting a C in math? I don't
believe it. You and Dad have a fit if we
get C's!"
Mom: "I guess we misunderstood how
hard one has to work for a C sometimes.
I'm sorry." (Good non-defensive reaction
and a diffused situation.)

Regardless of where you stopped in high
school, relearning math must occur in baby
steps. Starting at an earlier place is an
intelligent strategy. It may be months before
you are again math confident. Once in class, it
is important to accept learning math as a
mastery process. In its many layers, one learns
math as one peels an onion---uncovering one
truth at a time, one concept at a time, with
another layer of information awaiting
underneath.

Adult learners return to college with
divergent skills. Some have solid backgrounds.
For others, backgrounds resemble slices of Swiss
cheese --- gaping holes everywhere. All will
learn, but at different rates and to different
degrees of competency. I favor an affirmative
approach predicated on the knowledge that if the

core belief system is restructured, anyone can learn. By positively reprogramming the subconscious, one can be guaranteed success. That single effort will give one the courage to try and the stamina to persist.

SOMETHING TO THINK ABOUT

THE UNIVERSE WANTS YOU TO BE SUCCESSFUL. The Creative Spirit within you (God) wants you to pass math. The use of prayer, meditation, affirmations, or whatever you do to get centered, will help you in the empowerment process. Prior to beginning your studies, clear out your fears and self-doubts. You are not alone in this endeavor. Ask for Higher Guidance to help you do your work successfully, remembering to give thanks when you are done.

****

EXERCISES

1. Meditate for five minutes, DAILY, for one week, before doing any math homework.

2. Make three lists daily or weekly. (This technique is taught by Unity minister Catherine Ponder in her magnificent books on wealth and healing.)*

> List #1: What you wish to eliminate from your life.
> List #2: What you wish to bring into your life.
> List #3: What you have to be thankful for.

*

Ponder, Catherine. Open Your Mind to Prosperity. Unity Village: Unity Books, 1971

Reprinted with permission of Unity Books, 1901 NW Blue Parkway, Unity Village, MO 64065-0001

CHAPTER FIVE

AN ASIDE FOR PARENTS ONLY
OR THOSE WHO PLAN TO BE SOME DAY

SOME REFLECTIONS AND OBSERVATIONS

Sometimes I wish that it were possible to retrain all of America's bad math teachers, as poor teaching wreaks pain and havoc on psyches and futures. It is easier, however, to alter the approaches and perceptions of *students,* preparing them to learn defensively and overcome bad teaching. Although perfect teachers do not exist, there are a good many math teachers who perform empathically, humorously, creatively, incisively; who "drive home" the way to do it, giving students good feelings about math. So why have millions of Americans learned to dislike math?

Traditional education in America for years fostered a passive approach to learning. Sitting in rows, listening and writing, is how my generation and millions of younger adults went to school. Desired now is a more active, reactive, collaborative, affirmative approach. Students are getting more involved, speaking up more, injecting themselves into the process.

Mathematics cannot be learned passively, in silence. There are too many unanswered questions and incompletely described concepts. Furthermore, students come to a new course carrying baggage from previous encounters. Misplaced information, emotional scar tissue, disillusionment, and embarrassment can overshadow a first day in a new course. New beginnings are often weighted down with last year's wet blanket experiences. Students sit before the new teacher expecting to relive past traumas. What is needed is a way to overcome. This is the purpose of the affirmations in chapter one.

BELIEF SYSTEM

Belief system plays a major role in one's ability to do math. So called "blocks" are the accumulations of past traumas --- failure, humiliation, frustrated effort. Such blocks are frequently made worse by the fears and experiences of other family members. Group commiseration and identification, "I always hated math syndrome," all congeal to create a math family history of failure and struggle. These collaborative sharings somehow find their way into the "genetic predisposition" of the Joneses, the Williamses, and the Smiths.

PERSONAL ANECDOTE

My father was very good at math --- a major plus that outweighed my mother's dislike for it. But, when he told me that physics was hard, a poor high school physics teacher helped make it so. So here I was ---an anomaly --- a math major who could not do physics!

******

When our son, Patrick, at age 11, showed a serious predisposition toward astronomy, we bought him a brilliantly conceived space encyclopedia. Together, we explored quasars, pulsars, white dwarfs, and black holes. Recognizing his talent in science and his wish to become a space scientist, I carefully injected the word <u>physics</u> into the conversational equation. Not once, however, did I disclose my own dark past in physics. So he never knew the truth until recently. Now with a master's degree in nuclear physics among his many accomplishments, it no longer matters that Mom "hates the stuff." His predisposition to absorb the abstract is far superior to mine.

A larger part of protecting his future was the vigilance of his father and me to protect him from overzealous naysayers who did not believe in his potential, or had a problem with our son being an "A" student. I have spoken to many parents about the need to protect their children, while they are children, against discrimination of a different kind --- discrimination against bright children.

While this plight may seem bizarre, it does exist. There are teachers who have a problem with very bright children. There are also teachers who have a problem with very bright children of color, and the provocative light in which they place the false beliefs these teachers hold dear. My advice to parents is to guard your children from all sorts of negative classroom experiences as much as possible, until your children have the resilience to do it for themselves.

Protect your sons and daughters also from YOUR OWN math anxieties. Why bring your own fears and negative fantasies to life in front of them? If you hate math, hide it from them. *Do not proudly parade your failures unless you wish to see them duplicated.* Instead, tell them that math requires patience. Say that practicing math makes it easier. Explain gently that it is important to be neat, organized, and precise. Keep a ready supply of clean erasers, sharpened pencils, rulers, protractors, graph paper, and index cards in the house. Turn off the TV for two hours each night and sit with them, if necessary, while they do their homework. (Yes, you may read the newspaper or a book. Just be ready to assist if asked.) Teach them how to be assertive in a classroom. Tell them that math is more than numbers. That art is visual mathematics; that music is applied mathematics. Lay on the good stuff like good peanut butter --- thick and smooth.

ARE BOYS BETTER IN MATH?

Your answer may determine when you were last at a high school. Have you noticed that the math departments are 70% - 90% female? The same is true for many community colleges. You may wonder from where all these women have come! Previously, years of social engineering kept disproportionate numbers of women out of the physical sciences, math included, but that is changing.

Studies done in the 1970's revealed further that females were deliberately avoiding math dependent occupations while males were receiving much more attention in all classes. In fact, boys were being given more classroom time to "come up with" the answers as opposed to being told what the answer was. Girls, on the other hand, were being inadvertently coaxed to remain passive, by not being subjected to the dreaded classroom spotlight. Could this passivity have unintentionally influenced large numbers of women to avoid subjects requiring assertiveness, such as mathematics, physics, engineering, and chemistry? How much did it hurt women not to learn HOW TO LEARN DEFENSIVELY? By this I mean to take responsibility for one's learning as well as the credit for it. Learning to ask questions, seek additional help, insist on clarification, question unfair test practices, negotiate for higher grades are all necessary to survive the rigors of mathematics and all its bed mates.

PARENTAL REALISM AND SUCCESSFUL CHILDREN

One of the most frustrating experiences for any parent is to watch children flounder and underachieve. Having given them the best of our time, energy, and experience, we desire to see them aim high and get there on a sound trajectory. This does not always happen.

The present generation is restless and impatient. They grow up fast and somewhat recklessly. At best, we can help them identify their desires and plan strategies that coincide. We can provide opportunity. At the risk of shirking our "duty" we must ask, "Are our ambitions for our children consistent with their abilities and interests?" The child who would be a doctor must love science and exhibit the capacity to study/work for long hours. One could almost make a list.

Would be	Must be
actor	good at reading, memorizing
engineer	patient, a problem solver
dancer	strong, disciplined, creative

Obviously, children alone cannot make themselves successful. Needs must be met and potential nurtured. They must also be given the time to waddle through the ugly duckling stages of growth; the time to stretch, stumble, and build up their competence. (This is why it is so stupid for teachers to judge the *potential* of a 6th grader.)

What can be done to maximize potential and minimize the damage done in a public school setting? We can, as parents, assert the importance of a good education, and live up to our words by the environment we create at home, making it stimulating (intellectually, musically, artistically), and well organized (clean, well run, and reasonably quiet). We can limit the amount of stress absorbed at home. We can generate and maintain a harmonious, loving pasture for grazing, and offer guidance, space, freedom, and privacy. We can encourage our children to be creative and experimental. By introducing them to the performing and visual arts we can nurture the cognitive skills that are the underpinnings of academic success. We can build confidence by withholding our criticism. We can be the patient mentors who encourage success without demanding it.

I WANT MY CHILDREN TO BE GOOD IN MATH. WHAT SHOULD I DO?

From the beginning---keep a lid on your former negative math experiences. (1) Teach your children to count early and to think analytically: use money, puzzles, and games that teach one to think independently. (2) See that addition, subtraction, multiplication, and division facts are memorized by the end of second grade. (3) Set aside two non-negotiable hours of quiet time each day for reading, homework, test preparation, and piano practice. (NO, your spouse cannot watch TV in the family area during this time.) (4) Help your children with their homework to form excellent work habits of neatness and organization. (5) Run a well organized home. Can they find things? Can you? (6) Set your priorities wisely. Put the progress of your children ahead of community service and social activities. (7) Be available to them. Help them with school projects, frequently too complex for them to do alone.

(8) Keep an academic calendar that indicates when things are due, and insist your children record major events on it (reports, exams, science projects, etc.) (9) Make readiness for the annual science fair a *family* event. (10) Meet your children's teachers each year, and stay in communication with them, even in high school. (11) Remember that truly successful children succeed at everything because a success network is at their disposal, often beginning and ending with aggressively interactive parents. From an academic view, effective parenting means involved parenting, knowledgeable about every aspect of one's child's development.

****

EXCUSES. EXCUSES.

BUT, I WORK. So do I. That should be irrelevant.

MY CHILDREN DON'T WANT MY HELP. Have a Saturday family conference to calmly talk through issues, and make a plan that can work. One family member can mentor, another can tutor.

BUT, I CAN'T DO THEIR MATH. You can learn how if you take the time to read the book.

MY CHILDREN SAY THAT THE TEACHER DOESN'T DO IT MY WAY. Don't accept that. If the teacher's way were so perfect, why can't your child do it? It is more important to find a process that works than to get locked into a methodology one cannot understand or apply.

THEY SAY THE TEACHER IS BAD, DOES NOT ANSWER QUESTIONS, IS INTIMIDATING, MAKES THEM FEEL STUPID, ETC. ETC. Take off from work and go up for a conference together. (Single mothers should borrow a husband.) Get satisfaction or go higher.

I AM AFRAID THAT IF I PURSUE THAT I WILL MAKE IT BAD FOR MY CHILD. In that case, you are being cowardly, or afraid the teacher will be unprofessional. Take a stand. Defend your child's right to a good education.

MY CHILD KNOWS WHAT TO DO BUT DOES NOT DO IT, OR HE DOES HIS HOMEWORK BUT FORGETS TO TAKE IT TO SCHOOL. Either your child is over scheduled or has been allowed to become lazy and disorganized. If he is lazy, remedy the situation with motivation. Literally, find his "hot button" and push. Make him an offer, "do this or I'll snatch that!" If he is disorganized, help him to find "a place for everything" and to keep "everything in its place." If over scheduled, I ask again, are your ambitions for your child consistent with his energies, abilities, and interests? Are you living vicariously through your child without realizing that he is not you?

******

CHAPTER SIX

STARTING OVER ---
ANOTHER LOOK AT SOME BASICS

MATHEMATICS AND MUSIC

Had I not attended a high school comprised fifty percent of musicians, there would have been no basis for the beginning of this chapter. It was in the ninth grade at New York's prestigious High School of Music and Art that I decided to become a math teacher. Perched next to the windows of the fifth floor overlooking the City College marching band, my attention was sweetly divided between the words of a kindly, middle-aged math teacher and the brassy sounds of trumpets and horns floors below. A lovely memory converged into a career decision. Yes, I would do this. I would make math beautiful for students too.

A pianist from the age of seven, I learned the disciplines of reading music, counting rhythms, coordinating the eyes, hands, and feet, following sequential directions, integrating instructions (what to do) and restrictions (what not to do), appreciating the rigor of repetition and multiple memorizations, and performing in public (my greatest fear). All of the above indirectly prepared me to become a good student, in general, and a good math student in particular. Compared to mastering a Chopin etude, algebra and geometry were easy. As I watched countless musicians practice and perform for four years, I saw that persistence, flexibility, and patience were commonplace in our student population.

Years later, as a seasoned teacher, I listened to a learning disabilities teacher explain how difficult it is for learning disabled students to conceptualize mathematics, fractions in particular. As I listened, I remembered a cursory survey I take in my basic math classes each semester. I ask, "How many of you play a musical instrument?" Ninety percent of my basic math students usually answer NO, and each semester this startles me. In those classes in which students have a history of math weaknesses, there appears to have been very little formal musical training, and all of those significant learning skills that musicians know are missing! In my higher level classes (pre-calculus), I have found that up to 40% of the students are musicians. I do not believe that this discrepancy is coincidental! Every musician automatically does fractions and knows certain basics: $1 = 2/2 = 4/4 = 8/8$; knows that a musical measure comprises whole notes, half notes, quarter notes, eighth notes, multiple combinations, and much, much more.

Learning how to read music is an ABSTRACT skill, but no one *uses* that word, and music teachers assume that, with practice, the mind of a small child can learn to do complex things. We learn to associate the abstract (printed music) with corresponding sounds and rhythms, and we automatically assume that anyone who practices will learn to do this on any instrument. For most musicians, the pace at which the learning occurs is not a consideration, nor is the talent for learning. But, what musicians develop are the kinds of thinking patterns and study skills academics long to see in their math and science students: patience, willingness to practice, persistence, organization, ability to adapt to new situations (transposition), sight reading (playing what you have never seen before), problem solving expertise, and so forth.

Of course, not every child wishes to learn how to play the piano or any other musical instrument. However, due to the critical nature of the skills that learning music brings to the arena of learning, I would recommend some form of classical musical training for everyone. Just as helpful is formal training in voice or dance, as the latter teaches one to coordinate sight, sound, body, and rhythm. Extensive classical dance training (ballet, modern, lyrical, jazz) is far different from traditional sports training, and may generate the type of mental discipline that trained musicians acquire. Regardless of the vehicle, the study of music is the key.

In addition to learning how to play musical instruments, studies have indicated that young children who listen to classical music (baroque in particular) are more likely to develop a mind more acutely equipped for tackling the abstract later on. *

EXPANDING THE BASICS---
ADDITIONAL ISSUES FOR PARENTS AND
TEACHERS TO PONDER

Everyone learns basic number facts in elementary school. Unfortunately, for many of us, these are minimal and should be expanded to give us a greater flexibility and confidence when doing computations. A broader number facts basis should be memorized early on to include the multiples of 13: 13, 26, 39, 52, 65, 78, 91, 104, 117, 130, through the multiples of 19: 19, 38, 57, 76,...., 190. In addition, one should memorize the squares of the first 25 counting numbers, that is $1^2 = 1$, $2^2 = 4$, $3^2 = 9$, ..., $25^2 = 625$. Knowing these facts alone would facilitate arithmetic computations and make the learning of some aspects of algebra easier, namely the factoring of trinomials and the simplification of radicals.

A third inclusion would be the divisibility

* See page 112, regarding mathematics and music

rules, which are traditionally limited to knowing when a number is divisible by 2, 5, or 10. It is easy to remember the others. You know that a number is divisible by 2 when the last digit is divisible by 2. Did you know that a number is divisible by 4 when the last two digits of the number are divisible by 4? Or that a number is divisible by 8 when the last three digits form a number divisible by 8? In other words, the number 5489256 is divisible by 4, and we do not have to divide it to verify it. The rules for divisibility by 3 and 9 are even more fun. A number is divisible by three when the sum of its digits form a number divisible by 3 (The rule for nine is similar to the three rule). What is important is to remember that these or any body of facts makes the learning of future mathematics easier.

HOW MATH STUDENTS SHOOT THEMSELVES IN THE FOOT

Students tend to confuse opposite concepts. *They tend to say things backwards and think things backwards.* Because many concepts are described abstractly, it is important to write down illustrations of a definition in order for it to make sense when it is read again. Teachers and parents can help prevent students from reversing concepts. For instance, some students confuse the concept of squaring a number with taking the square root of a number because they do not make a visual association with the verbal definition and because they confuse the "direction" in which they are mentally processing. Let me illustrate.

Squaring 8 means multiplying $(8)(8) = 64$. Taking the square root of 8 means looking for a number x, such that $(X)(X) = 8$. The number x is not going to be found except in a table of square roots, or on a calculator, because it is a non-ending, non-repeating decimal we call an irrational number. This brings us back to that list of perfect squares that students should memorize. Only certain positive numbers have easy to identify square roots. These special numbers called perfect squares are 1, 4, 9, 16, 25, 36, 49, 64, 81, 100, 121, 144, 169, 196, 225, and so forth. All of the other whole numbers not found in this list have irrational square roots. So, whereas the square root of 25 is 5, the square root is 24 is irrational, but lies between 4 and 5, closer to five.

Many students confuse unrelated concepts such as "even" with "positive " and "odd" with "negative." When descriptions in algebra use dual adjective definitions such as "a is a positive real number..." students should be encouraged to form a visual frame of reference for what is being described, as opposed to depending on the words themselves. For instance, it is wise to use a number line to visualize "positive, real number."

The real number line

$$\underline{f(x)} \quad \longleftarrow\!\!\!+\!\!+\!\!+\!\!+\!\!|\!\!+\!\!+\!\!+\!\!+\!\!\longrightarrow$$
$$\qquad\qquad -5\ \text{-}4\ \text{-}3\ \text{-}2\ \text{-}1\ \ 0\ \ 1\ \ 2\ \ 3\ \ 4\ \ 5$$

****

CHAPTER SEVEN

BEING THE RIGHT TEACHER
(A CHAPTER JUST FOR TEACHERS)

I am troubled by several trends I see in the teaching of math. First, the textbooks are becoming needlessly difficult. With the filtering down of material and concepts, more content is pushed downward into courses with little consideration for the ability of students to understand it. Material that once was reserved for high school, is now found in the curricula of middle school. Another example of this would be the utilization of very complex equations in a pre-calculus course to illustrate a simple concept like a quadratic equation simply because the implementation of graphic calculators makes their solution possible. Whereas the equation $2x^2 + 5x + 6 = 0$ might have been the original model, it may now appear in the form $2.87x^2 + 0.987x - 0.651 = 0$, which can be done only with a calculator.

Second, concerned about "weeding students" from overcrowded classes, some teachers are deliberately distancing themselves from their students, and many students are finding themselves in a hostile environment. Even those instructors who are kind and compassionate must cover more material, use more technology, and teach larger classes, particularly at the college level. Universities do not seem to care how much math students can learn in lectures containing 80 students or more.

Third, no one wants to admit that "these books are too difficult, these classes are too big, or there just is not enough time to cover the content of this syllabus." A game of compromise ensues and students loose. At the college level the remedy is to take the money and flunk them when it does not work.

With the above commonplace, technological reform and attention to alternative learning styles *alone* are not going to make people succeed in or like math more. At best, even as we encourage innovative or collaborative learning, many students continue to feel too confused and insecure to benefit from new adaptations. Furthermore, college level students are not the only ones being disadvantaged.

STUDENT COMPLAINTS

Over the years my husband and I have tutored many students who were literally suffering in their middle and high school math classes. As I list their most frequent complaints, I will try to place these in proper context.

1.   *"The teacher goes too fast."*
     -too much material to cover
     -class is too intimidated to ask questions
     -insensitivity on the part of the teacher

2. *"We are not following the book."*
    -poor choice of text (not easy to read and
        not logically sequenced)
    -teacher is disorganized
    -teaching style of teacher fails to clarify
        procedures for everybody
    -teacher does not create a sense of continuity
    - teacher has not provided the class with a
        syllabus.

3. *"The class is too noisy. I'm not learning
    anything."*
    -teacher exerts poor classroom control
    -teacher funnels his attention toward a
        chosen few, ignoring the majority of
        the class
    -group work is poorly defined and poorly
        orchestrated
    -basic fundamentals or the chapter at
        hand has been incompletely explained;
        hence, no one knows what to do.

4. *"The teacher does not answer our
    questions."*
    -teacher is preoccupied with getting from
        point A to point B
    -students are not being encouraged to ask
        questions

-teacher is unaware of how to solicit
   questions without making students feel
   stupid (ie., teacher has been poorly
   trained)
-teacher is adhering to the philosophy
   "We 've already covered that and once
   is enough."

In a word, HELP!

DEFINING THE GOOD TEACHER

Every teacher of mathematics recalls how
difficult it was in college to read the text and
bridge the gaps between statements. It was a
"joke" beyond calculus, that writers, responding
to publisher demands to condense text material,
would say that a conclusion was obvious. As
college students we tolerated these lies. (What
choice did we have?) It became our badge of
competency to be able to extrapolate omitted
information, no matter how lengthy or tedious
the process. Students in middle school or high
school should not have to endure these trials. (It
is important that teachers select texts that are
both easy to read and logically sequenced. It is

important that publishers be made to understand that textbooks which lack complete explanations are essentially valueless to students, and the best texts will allow an absent student to keep up on his own.)

Unfortunately, being forced to self-teach has created cynicism among *unsympathetic* teachers. *Some* teachers feel that students who cannot also self-teach don't deserve to take math.   This attitude has been a blight on the teaching profession.

As difficult as learning math has been for many math majors, there is an important lesson here: *good teaching is empathetic.* Good teaching does not imitate the book in its need for brevity. Good teaching does not assume that things are obvious. The good teacher does not believe that less is more, and does not ramble on, trivializing new material, aggravating wounds, dampening spirits, bruising egos, and crushing dreams.

The good teacher does not disparage student ability, recognizing that even the best students experience academic dry spells. She/He understands that teens especially are at high risk for inexplicable performance lulls, because they are more vulnerable to failure in times of personal crises. Unlike adults, who have learned to "grin and bear it," to hide personal pain beneath professional persona, teens may collapse in one area in an effort to cope in another. *The good teacher seeks out the reasons for poor performance and offers empathy and remedies, realizing that ability is often not the issue.*

In seeking out good teachers for my own children, it became increasingly important that other traits superceded good credentials or job longevity. It was more important to me that my children have teachers who were fair, caring, patient, and encouraging. It was critical to me that they avoid teachers known to be discouraging, abrasive, thoughtless, or lazy.

At my community college, in interviewing prospective math teachers, we assess personal philosophy and personality before examining teaching style. (Yes, we see how they teach.) From a pool of highly qualified applicants, we look for fairness, integrity, enthusiasm, a high level of energy, and a commitment to diversity. We understand the need to be clear and structured, and to generously explain concepts. As a result, we comprise a rare environment----

a department containing many master teachers.

In my opinion, good teaching, far from being an individual effort, is a spiritual duet between oneself and one's Creator. One prays for guidance and God delivers serenity, harmony, clarity, and purpose. The interaction between the student and teacher evolves into a unique blend of respect and trust. As a vehicle for divine intervention, the good teacher has the patience to transform a subject feared and hated into a platform for possibility. The classroom becomes a place where people can learn, laugh, experience joy and freedom, find acceptance and friendship. The good teacher appreciates the opportunity to share, captivate, perform and love, drawing a crowd of wary spectators into the fascinating world of mathematics.

****

EXERCISES

SOME THINGS FOR TEACHERS TO THINK ABOUT

1. How do you know if you are on target?
2. Are your students excited about what they are doing?
3. Is there genuine laughter in your room?
4. Do you sense that usually you have their undivided attention?
5. Are your students relaxed?
6. What is their response to your questions? Do they answer willingly or "shut down"?
7. Do they work well together collaboratively?
8. Do your students trust you?
9. Does knowing the nature of their problems make you empathetic or cynical?
10. As your students exit the classroom, do you ever hear a "thank you"?
11. Are you systematically making concepts easy to grasp?
12. Are you willing to go over material a second or third time if you sense they need it?
13. Do you answer questions thoroughly and patiently?

14. Do you format complete solutions on the board at all times?
15. Do your students compliment you to *your* peers and recommend you to theirs?

CREATING PARADIGM SHIFTS

Please rank order your position on the following statements.
(1: I do not agree.   5: I concur completely)

1. I am interested in mentoring my students.
2. I believe each student needs someone to believe in him (her).
3. I believe that everyone wants to succeed.
4. I believe that each student can learn how to get A's.
5. I believe that there is nothing elitist about learning math.
6. I believe that learning math can be a joyous experience.
7. I believe that my students have more intellectual and creative potential than I will be able ever to assess.
8. I believe that my teaching has room for improvement in both style and pedagogy, and I am open to suggestions.

CHAPTER EIGHT

IT'S UP TO YOU NOW
*(SOME CLOSING PHILOSOPHICAL THREADS FOR
STUDENTS)*

I am a great procrastinator. I tend to wait until all circumstances are "perfect" before engaging in a choice that is truly good for me. As a consequence, I have missed a great many opportunities to empower myself with life-enriching activities. I suppose what has held me captive is fear, plain and uncomplicated. The fear is that I am going to make a mistake, be disillusioned, waste my money, misspend my time, and not meet with the approval of others. Does any of this sound familiar?

Eventually, we come to know that there are several voices which speak to us. (Unlike the vision many of us bring from childhood, there are *not* just two voices, one good and one bad). These voices take many forms, and the voice that deceives us into bad decisions, or the procrastination of good decisions, pretends always to be on our side.

As we become wiser risk takers, we stop hedging and trust ourselves to leap. Some net will catch us, for as every artist knows, a first attempt will not yield a masterpiece. In attempting a new skill or returning to an old one, we will not necessarily lead the pack. As students we must remember we are not in class to compete to be the best, the brightest, or the first to finish. Our desires have changed. We want to face past demons and trash them. We want to get on with our agendas so we can be jubilant about our lives. We must be willing to take that first step, whatever it is.

I wish to commend all of my "older students" for setting aside their fears and signing up for my course. I could be boring you and weighing down your schedules with relentless assignments. But, you had the vision to take a chance on yourselves, with me, in math. CONGRATULATIONS! YOU ARE HERE! Once the decision is made to sign up for a math course, please remember that a skill is being tested which has been dormant. Like the dancer who has not been at the barre for ten years, or the pianist who dusts the piano but does not play it, you have become rusty. Because you are out of practice, the return to peak performance may be a difficult climb. Hours of study may appear to yield nothing. DON'T QUIT! You are training the mind to reinvent new patterns of thinking. The breakthrough will come. Like the pianist whose fingers have "memorized the work," your mind will need to practice tasks over and over until it trusts itself to "play on its own."

Just as writing is an eye-hand-brain collaborative event, so is the execution of mathematics. The channels to correct computation and speed must be re-excavated and reinforced. New words, old phrases, fresh understandings, and clarified procedures must all be reconnected in the brain. A new matrix is being constructed, linking language with visual imagery, symbolic summaries to theoretical concepts. Accomplishing all these tasks takes patience. No quick fixes can mend the present.

This is why the affirmations are so important. They will give you the courage to persist. With a better framework in which to frame your image of yourself and your capabilities, you will succeed. Past hindrances will tumble away. YOU *CAN* DO MATHEMATICS. YOU *WILL* DO IT.

EXERCISE

Go back to chapter one. Select seven of your favorite affirmations and write them over seven times. Do one each day for seven days.

MOVING BEYOND THE BIAS

Mathematics is the singular subject, in my experience, that must always justify its existence in the minds of people. Always asking "Why must we learn this?" students are vocalizing strong objections to a subject they have come to despise and fear. The issue of justification does not appear to surface as much for other disciplines. Did we really have to memorize every battle and treaty detail in European history, the skeletal structure of plants and insects in biology, or the poetic forms of the nineteenth century? Probably not. But, most of us did so anyway without a fight.

Learning should be about stretching, not stuffing. Learning math forces both brain hemispheres to work together, teaching us to reason and to conceptualize. Math prepares us to do biology, chemistry, accounting, business, English, and art, by conditioning us to organize, analyze, and synthesize information. It will lead us to witness reality on a higher, symbolic plane. Other leaps will manifest. *As the bridge between the dimensions of body and spirit, mathematics is the conduit which will yield us the answers we have not yet accessed. The answer to the ultimate mystery, "Who are we?" will be answered by physicists and found in mathematics.*

******

BREAKING THROUGH BARRIERS

WHAT THE WORLD CALLS SUCCESS MAY NOT BE SUCCESS FOR US AT ALL. We have to define it for ourselves and get in step with our personal delineation of it. In my opinion, to have a healthy and beautiful self-concept is to be successful. To have a strong sense of personal fulfillment, to lead a joy-filled life true to yourself without outer harassment or inner turmoil, and to live in complete serenity is success. *Success is self-actualization* --- being all you can be, realizing that for which you were put here. We are the only ones who can give that identification meaning.

We need to be honest about what is important to us, separating our vision from the energies of those who have defined it for us in the past. Men need to define themselves as themselves and for themselves. Women need to break through societal barriers that constrain and limit the definition of womanhood. We need to see ourselves as spiritual beings, not as commodities with genders for commercial purposes. We must find the courage to cast away our illusions. We do not need them to be happy.

NEW BELIEF SYSTEMS

As we rid ourselves of cynicism and negative thinking, we allow ourselves the opportunity to dream. We create paradigms of hope and believe that new things are possible for us. We construct new agendas and generate new opportunities. Our mind's eye opens and choices become clear. We free ourselves for greater and fuller accomplishments.

It is important that we construct new lives as well as new belief systems. If we have been too busy, that must change. We cannot maximize our potential living on the "fast track." Speeding through life generates only the *illusion* of success. Becoming who we are means to slow down and see. It means to reexamine our agendas, and carefully discern what is important and act upon that recognition. It means to put ourselves first, to calm our minds, to reign in our emotions, to use our energy to empower ourselves. This is how we become successful.

******

SURVIVAL
(How life brings us the unexpected)

Just as we get our feet on earth
The earth moves
As soon as we get comfortable in our lives
Our lives erupt
We toss our lot toward the familiar
and it changes hue
We find ourselves in hostile lands
Upon which nothing seems to smile at us
But, we survive
As we learn to face what's different
without fear
and, with love.

REALITY

Success is born when we choose to dismiss the past and live in the future we make today. We block ourselves by getting stuck on issues over which we have no control: the problems of others, the whims of family, or the desires of those we love. We must arrive at a position of strength and indifference. Regardless of how others choose to see us or choose their own circumstances, <u>we must move forward.</u> We cannot let ourselves wallow in their turmoil because of our empathetic/sympathetic natures. We must give them the right to be themselves, and we must separate ourselves from their identities. This does not mean we cease to love them. It is just the realization that we cannot incorporate a disproportionate amount of their energy into our lives. WE ARE NOT THEM. Their issues, goals, needs, and desires are not ours. As we seek out the truth for ourselves, we make the right choices. We begin to incorporate an agenda that works for us. We become true to who *we* are. We find our dharma, our purpose for living. We discover serenity, joy, and self-fulfillment. We are freed to realize accomplishment.

WHAT HAPPENS AS WE GET IN STEP WITH WHO WE ARE?

We experience
peace of mind---better health
greater focus and vision
better use of time and energy
achievement and personal success
fulfillment---joy and happiness.

WHAT CHARACTERISTICS MAKE ONE GOOD AT LEARNING?

Stillness
The ability to focus
Concentration-- to draw in information
Reaching--inward and upward
Facility in writing
Organization
The ability to focus on long range goals
      and surrender short term gratification
Intuition-- the ability to perceive the
      moods of others
Confidence/Competence --knowing one's
      basic skills are intact
Knowing oneself--appreciating one's
      individuality

GETTING PAST THE BAGGAGE

Life is meant to be an adventure, a thrust
into creative living, a taking of risks, a chancing
of outcomes, a joy. But, we get caught into webs
of circumstances: bills, constraints, and
stressors (like traffic and bosses). We lose sight

of the goal and forget to enjoy the journey. Joy gets transmuted into drudgery. What happens to us? Do we allow reality to close in upon us? Or, are we unrealistic in terms of what we dream possible? If we choose to be as stress free as possible, we must set limits on how much activity we engage in and how much money we spend. We must choose parameters for our behavior that allow us to keep to our success schedules and to be content. We must remember that *our time is limited.* Making others happy at our own expense throws us off balance by making us stretch inappropriately. Saying NO frees us to maintain the agendas we set. Saying NO keeps the desires, needs, and agendas of others from taking over our lives.

Once a commitment to ourselves has been made, step two is to find outlets for joy as well as success. The right study/working environment (quiet, musical, sunny, colorful, book-filled)---whatever we need, should be established.

Next, a schedule should be created that allows our studying to be the priority. As difficult as this sounds to some, UNLESS THIS DECISION IS MADE the effort to complete a class and ultimately obtain a degree will be very, very difficult. If the priorities are wrong, tests will be failed, courses dropped, and graduations postponed. We will rebound from issue to issue trying to justify why we should come first. Think of the thousands of students who have overcome personal and math anxiety long enough to get their diplomas. Not selfish, these people are just centered, in the middle of their own stages.

LIFE IS WONDERFUL---A NEW OUTLOOK

Like a valley overrun with wild flowers, life abounds with opportunities for our growth and advancement, through which we can run and choose what we like. We need to get past fear--- to excise it, to pluck it out like a splinter. These lives belong to us. As we find the courage to claim them, we will succeed at everything: math, writing, painting, singing, skiing, sky- diving, and law. We will see all of life in the proper sequential progression. Like mathematics, every accomplishment will be built upon a succession of small steps which collectively become big leaps. Let us affirm our right to be free. Let us affirm our right to be successful. We are magnificent creatures.

Let us remember that our abilities are God-given and God-driven. The infinite power of the Creation that indwells each of us makes all things possible. Many of our conflicts arise, because, in addition to forgetting the above, we fail to acknowledge that an unseen hand guides our destinies. More lives than ours are involved in our progression. We get too outcome oriented. That is why we are always fixated on the final results and also why we cannot have them now. Life is a process, as is accomplishment. We need to learn to enjoy the journey.

CONCLUSION

As ambitious people, we want success at different levels ---personal, professional, and creative. If we visualize these aspects of our lives as sine curves, we find that they are usually out of sync (phase) with each other. Occasionally, there are those rare and beautiful intervals in our lives where all three aspects are simultaneously increasing or peak together, and we are radiantly happy.

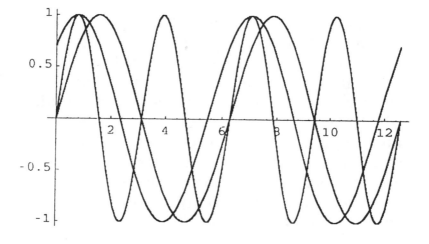

My wish for you is that you will find the courage to rethink what you have believed and the choices you have made, and move aside the barriers that have stood before you. That you will incorporate into your life affirmative means for self-empowerment. That you will reach the summit of your career and experience greater heights of personal fulfillment. Most of all, I wish you joy as you experience the journey. God bless and Godspeed.

NOTES

Sandra L. Manigault

Notes

*

    Researchers at the University of California, at Irvine, have shown that exposure to formal musical training increases the ability of preschoolers to reason and learn. Frances Rauscher and Gordon Shaw administered tests of spatial reasoning to preschoolers who were given piano and voice lessons by professionals at the university over a period of eight months. These children dramatically outperformed the control group who were not given daily music lessons.[1]

    At the University of Dusseldorf in Germany, neurologists Gottfried Schlaug and Helmuth Steinmetz, indicate that early exposure to musical training appears to positively impact the brain of a young child by reinforcing neural connections and possibly creating new ones.[2]

[1]   "Tuning Up Young Brains." *Science News*, Vol. 146, no. 9, (August 27, 1994), p. 143.

[2]   "Music of the Hemispheres." *Discover*, Vol. 15, no. 3, (March 1994), p. 15.

ORDER FORM

The Book For Math Empowerment may be ordered through the mail for $ 12.95 plus postage and handling. Tax is required only if the book is to be shipped to a Virginia address.

(Please Print)

Name _____

Address_____

City, State, Zip Code _____

Telephone Number (_____)_____

copies ordered _____

4.5% sales tax (Va. residents only) _____

Include $3.00 (postage/handling) for the first book; $ 1.75 for each additional book _____

Total enclosed _____

Please make all checks or money orders payable to Godosan Publications and mail to

    Godosan Publications/ S. Manigault
    P.O. Box 3267
    Stafford, Virginia 22555
    (540) 720-0861   Phone and Fax

Please inquire about our special discount rate for large orders. Please allow 3-4 weeks for delivery. Thank you for your order.